Health 126

快速止痛药

Fast Relief

Gunter Pauli

[比] 冈特·鲍利 著

[哥伦] 凯瑟琳娜·巴赫 绘

章里西 译

上海远东出版社

丛书编委会

主　任：田成川

副主任：闫世东　林　玉

委　员：李原原　祝真旭　曾红鹰　靳增江　史国鹏
　　　　梁雅丽　孟小红　郑循如　陈　卫　任泽林
　　　　薛　梅　朱智翔　柳志清　冯　缨　齐晓江
　　　　朱习文　毕春萍　彭　勇

特别感谢以下热心人士对童书工作的支持：

匡志强　宋小华　解　东　厉　云　李　婧　庞英元
李　阳　梁婧婧　刘　丹　冯家宝　熊彩虹　罗淑怡
旷　婉　王靖雯　廖清州　王怡然　王　征　邵　杰
陈强林　陈　果　罗　佳　闫　艳　谢　露　张修博
陈梦竹　刘　灿　李　丹　郭　雯　戴　虹

目录

Contents

一头驴受到一阵阵头痛的侵袭，感到不知所措。他的朋友山羊刚好过来，注意到了他脸上难过的表情。

"看起来你昨晚一夜未眠哪！"山羊说，"是不是昨天你为我们守夜的时候狐狸和草原狼让你疲于应付哇？"

A donkey is suffering from a headache. His head is pounding and he does not know what to do. His friend, the goat, comes along and notices his sad face.

"It looks like you had a sleepless night!" says Goat. "Did the foxes and coyotes keep you up last night when you were guarding us?"

一头驴受到一阵阵头痛的侵袭。

A donkey is suffering from a headache.

......把姜配上大量的糖一起吃......

...eat ginger, mixed with lots of sugar ...

"可不止昨晚一个不眠之夜呀！我这个头痛实在是太严重了，弄得我白天休息不了，晚上也睡不着。"驴解释道。

"你应该试试把姜配上大量的糖一起吃。"山羊建议道，"吃完以后你的头痛应该很快就能康复了。"

"Many sleepless nights! My headache is so bad, it leaves me no time to relax during the day or rest at night," Donkey complains.

"Well, you should try eating ginger, mixed with lots of sugar," Goat suggests. "That should do the trick, and get you on the pathway to recovery right away."

"糖和姜一起吃？"

"对，姜可以舒缓神经，而糖可以帮助姜的成分尽快达到大脑。"

"Sugar with ginger?"

"Yes, the ginger eases the nerves and sugar will help get it to your brain fast."

糖和姜一起吃？

Sugar with ginger?

……糖只会危害我的牙齿健康。

... sugar will only relieve me of my healthy teeth.

"用糖来快速止痛吗？我觉得你弄错了——大量的糖只会危害我的牙齿健康。"

"相信我。你的大脑并不能储存任何能量，所以它需要糖。"

"Fast relief because of sugar? I think you got it wrong – lots of sugar will only fast relieve me of my healthy teeth."
"Believe me, your brain does not store any energy, so sugar is needed."

"我原以为肌肉才需要糖以便更好地工作。"

"没错，糖供应能量。但当它来到大脑的时候，它还能帮助运送氧气。"

"哦，我明白了。"驴说，"如果我脑部能有更多的氧气供应，疼痛也能快速减轻了。"

"I thought it is my muscles that need sugar to work better."
"Sugar provides energy, but when it gets to the brain it helps transport oxygen."
"Oh, now I get it," Donkey says. "If there is more oxygen available in my brain then it gives me fast relief from pain."

糖供应能量，还能运送氧气。

O₂

Sugar provides energy and transports oxygen.

山羊，你真是个不错的朋友。

Goat, you are a true friend.

"没错，糖就能达到这个效果。"

"谢谢你，山羊，你真是个不错的朋友。我会把糖的好处跟我的兄弟姐妹宣传的。"

"但他们住得很远吧。"

"Indeed, and the sugar will help to get the job done."
"Thank you, Goat, you are a true friend. I will tell my brothers and sisters that sugar is not all that bad after all."
"But they live miles and miles away."

"你不知道吗？驴可以隔着上百千米的距离呼唤对方。"

　　"不需要用电话吗？"

　　"当然。别忘了，我们跟黑猩猩一样也是群居生活。我们是个有爱的大家庭。我们会时刻用听觉关注家里其他成员在哪里。"

"Don't you know that donkeys can call each other over very long distances, even from over a hundred kilometres away?"

"Without using a phone?"

"Of course. And don't forget, like chimps we like to live together. We love to be a close family. We always have an ear out to hear where the others are."

......用听觉关注家里其他成员在哪里。

... have an ear out to hear where the others are.

……长那么长的耳朵干吗？

... why do you have such long ears?

"这么一想，你们耳朵这么长，就是为了能听得更清楚吧？"

　　"我们听力并不比马更好，但我们更健壮，记忆力也很好。"

　　"既然不是为了能让你们听得更清楚，那长那么长的耳朵干吗？"

"I wondered why you have such long ears. Is it to hear better?"

"We don't hear better than horses do, but we are stronger than they are and remember everything very well."

"So, if it is not to hear better, then why do you have such long ears?"

"别忘了我们是在沙漠里出生的，我们的长耳朵能帮我们保持凉爽。"

"你真是头很酷的驴。"山羊的眼睛里泛着光。

……这仅仅是开始！……

"Remember we were born in the desert, and our long ears help to keep us cool."

"You are rather a cool donkey," Goat says, with a twinkle in her eyes.

... AND IT HAS ONLY JUST BEGUN! ...

......这仅仅是开始！......

... AND IT HAS ONLY JUST BEGUN! ...

虽然驴的体型和马很相近，但它们其实比马更强壮。驴的记忆力很好，能认出 25 年前遇到的人或者路过的地方。驴能听到 100 千米外的同伴发出的呼唤。

Donkeys are stronger than horses, even though they are almost the same size. A donkey has a very good memory, and can recognise people and places 25 years later. Donkeys can hear the call of other donkeys 100 km away.

驴是群居动物，会像猴子、黑猩猩这些动物一样梳理同伴的毛发。驴以个性固执闻名，但事实上它们只是不会因为强迫或者恐吓就去做自己觉得危险的事情罢了。

Donkeys are herd animals and will groom each other in the same way that monkeys and chimps do. They have a reputation for stubbornness, but that is because they will not be forced or frightened into doing something they consider unsafe.

The female donkey is called a Jenny, the male is called a Jack and their young are called foals. Donkeys live for up to 50 years. Their hoofs are so tough that they do not need horseshoes.

母驴有个英文昵称叫 Jenny，公驴的昵称则是 Jack，小驴驹的英文叫 foal。驴最长能活 50 年。它们的蹄子非常结实，不需要戴蹄铁。

Ginger has been used for at least 2,000 years as a natural remedy for nausea, diarrhoea, upset stomach, to aid digestion and to combat headaches, including migraines.

姜被用作天然药物已经有 2 000 年的历史，主治恶心、腹泻、腹痛，还可帮助消化，治疗包括偏头痛在内的各类头痛。

Ginger oil contains more than 200 different substances. Ginger can either be ingested as a juice, inhaled as a vapour when adding the oil to hot or boiling water, or applied as ginger paste on the forehead.

姜油中含有超过 200 种不同的成分。可以把姜榨成汁液服用，也可以把姜油倒进滚热的水混合蒸汽吸入，或者制成姜膏抹在额头上。

$$CH_2 - CH - CH - CH - CH - CHO$$
$$\quad\ |\qquad |\qquad |\qquad |\qquad |$$
$$\quad OH\quad OH\quad OH\quad OH\quad OH$$

葡萄糖

Brain cells cannot store energy and therefore require a steady flow of glucose. Glucose increases the blood's ability to transport oxygen to the brain, and this effects the relief of a headache.

脑细胞不能储存能量，所以需要持续的葡萄糖供应。葡萄糖能提高血液运送氧气到脑部的能力，这有助于缓解头痛。

糖类可用来预防腹泻的儿童出现脱水。我们吃的糖在口腔中会变为酸性，破坏牙釉质的保护功能，引发细菌感染，导致蛀牙。

Sugar is used to prevent dehydration in children who suffer from diarrhoea. The sugars we eat turn acidic in our mouths and destroy the enamel, the protective layer of our teeth, allowing bacterial infections to cause cavities.

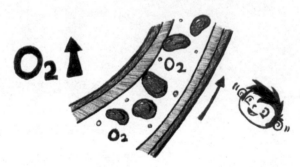

血液中氧含量的增加可以中和身体产生的毒性物质和酸性废物，淋巴系统将代谢废物从组织中排出，可减轻压力，缓解肌肉紧张，从而达到止痛的作用。

Increased oxygen in the blood neutralises toxic and acid waste from the body, moves fluid out of the tissue through the lymphatic system, and reduces stress and muscle tension. In doing so it eases pain.

When you have a headache would you prefer to take a pill, use deep breathing for a few minutes or take some ginger with sugar?

当你出现头痛的时候，你是更倾向于吃止痛片，深呼吸几分钟，还是吃拌了糖的姜？

Would you be able to remember a road you once travelled on 25 years ago? What would you call someone who can still remember something like that after so many years?

你能记得住25年前走过的一条路吗？对于那些能记住如此久远事物的人，你会如何称赞他？

Do you believe that donkeys are simply stubborn, or that they are just being careful when they do not want to move because they sense a risk?

你觉得驴就是固执，还是它们只是在感受到危险时保持谨慎所以不肯挪动？

How can long ears help a donkey cool down in the desert?

驴的长耳朵是如何帮助它在沙漠里保持凉爽的？

Do It Yourself!

自己动手！

Next time you have pain consider the treatment options you have. You can take medicine that will soothe your pain, but this will take time. Or you could start breathing calmly and deeply for a few minutes. Ask some friends to do the same and compare results. Which treatment works the fastest and has the most reliable outcome?

下次你感到疼痛的时候，想想你可选的治疗方法。你可以吃药止痛，但这需要一段时间才能见效。或者你可以开始慢慢做几分钟深呼吸。让几个朋友跟你一起这么做，比较一下治疗的效果。哪种治疗起效最快并且最为可靠？

学科知识

Academic Knowledge

生物学	焦虑、愤怒、忧虑、悲伤、抑郁这些情绪是失眠最常见的诱因；头痛是由脑活动过度和情绪紧张导致的；疼痛是类似于嗅觉、触觉、听觉的一种知觉；神经细胞将疼痛的信息传递给脊髓，随后再送至大脑；受体蛋白可以感知不同类型的疼痛；姜是一种开花植物，其根部可以当作辣味调料或者入药；姜和姜黄以及小豆蔻同属一科；血液在肺部与氧结合；驴是一种理想的看护羊、牛等牲畜的动物，它对狼和狐狸等犬科动物有天生的敌意，会用脚踢和撕咬来抵御来犯。
化 学	姜辣素是姜含有的一种重要化合物，它和辣椒素以及胡椒碱是"近亲"；姜烯是姜最主要的成分，占到了姜根精油的30%；在中世纪姜被用于制糖。
物 理	声波传递的速度与传播介质的惯性及弹性系数有关；弹性系数指的是物质在外力作用消除后恢复原有形态的倾向性。
工程学	声波是通过粒子间相互作用在介质中传播的一种压力振动；波传导的速度取决于传导的介质。
经济学	在14世纪，一千克姜和两只活羊的价格相当；全球止痛药市场的规模超过1 000亿美元；自从美国科罗拉多州颁发大麻零售许可证后，大麻种植已经取代石油天然气工业成了当地经济的支柱产业。
伦理学	过量服用止痛药可能会导致患者成瘾，虽然这可能会让开处方的人在经济上获益，但会严重影响患者的健康。
历 史	驴至少在公元前2800年左右就已经在埃及被驯化；姜在早期从原产地印度出口至古罗马。
地 理	驴来自非洲。
数 学	速度等于距离除以时间。
生活方式	睡眠缺乏和过度睡眠均可能引发偏头痛；而家庭带来的幸福和舒适可以缓解头痛。
社会学	在英语中，驴的叫声常被拟声为"hee-haw"或"eeyore"，小熊维尼中小驴的名字便来自于此；驴重友情，记忆力很好，寿命可达50年；西方国家的人用"锐痛""跳痛""刺痛"这些词来形容疼痛；其他文化会用闪电、根系复杂的树木、蜘蛛网、鼓和笛子的声调这些天然元素来解释疼痛；日本文化中的"忍"强调隐藏疼痛和情感；亚洲国家还有包括针灸、芳香疗法在内的多种天然止痛方法。
心理学	疼痛并不是由组织损伤继发产生的电冲动，而是当神经信息输入后人所体会到的一种情绪感受；短时间的疼痛叫作急性疼痛，长时间的疼痛则叫作慢性疼痛；人在经历过剧痛后，就不会再被剧痛所击倒了。
系统论	虽然慢性疼痛的诱因可能比较单一，但它可以继发许多其他效应；疼痛的程度和原因通常取决于一些难以控制的心理因素。

情感智慧
Emotional Intelligence

山羊

山羊观察力敏锐，对驴的头痛表现出了关切，因为驴是其族群的守护者，可以保护他们免受犬科动物的袭击。她所给出的治疗手段让驴感到很惊奇，但山羊随后就解释了为何姜和糖的组合可以奏效。山羊非常有耐心，按逻辑顺序一步步向驴解释背后的细节，直到驴确认自己已经完全理解。当驴热切地想把这种新疗法分享给自己的亲友时，山羊思考驴究竟是通过什么方式在没有电话的情况下跟亲友通信。山羊在问驴一些私人问题（如为什么他的耳朵这么长）的时候表现得很从容。在驴没有立刻给出回答的情况下，山羊不断追问，并表现出了对驴的共情。

驴

驴抱怨自己好几个晚上睡不好觉，白天也因为头痛无法休息。他承认自己不知道姜和糖混合的偏方，但他信任山羊的说法，为了了解更多细节接连问了几个问题，以完全弄清偏方背后的玄机。他明白了之后，便想把这些信息立刻分享给他的亲友。驴非常自信，知道他可以轻松地把消息传达给远方的族人。他告诉山羊，他的亲眷们是一个紧密的大家庭，他的家人也会随时侧耳聆听他从远方传来的讯息。他知道自己很强壮，把自己跟马进行对比，还跟他的山羊朋友分享了耳朵的秘密。

艺术
The Arts

试试看能不能模仿驴的叫声。要想尽可能让自己的叫声逼真，你得试着扯开自己的声带，重点模拟发出和驴的叫声接近的音符。驴一般通过喊叫传达痛苦的信号，所以试着通过我们的声音表达焦虑痛苦的感情——仿佛我们就是那头感到危险的驴一样。

思维拓展
Systems: Making the Connections

疼痛是个很难解释的概念，每个人、每种文化对于疼痛的理解都不尽相同。急性疼痛往往会让我们迅速做出反应，而慢性疼痛的成因常常超出我们的理解范畴，其治疗也更难入手。由于止痛药可能成瘾以及带来不良反应，越来越多的患者开始寻找止痛药以外的减缓疼痛的方法。像针灸、芳香疗法、催眠这些疗法已经被历史反复检验确实具有效果，还可以避免化学物质依赖的风险。但许多慢性疼痛是由现代生活方式导致的，譬如缺乏锻炼，久坐，摄入酸性食物过多，将糖类作为主要能量来源，大气、水和食物的污染导致毒性成分在体内累积等。我们现在正逐渐与我们祖先的智慧背道而驰：水果蔬菜、块茎类食物、草药以及香料中蕴藏着很多治愈的能量，吃这些新鲜的食物可以让我们获得许多营养成分，保证我们拥有强健的体魄和思想，以及抵御疼痛的能力。

动手能力
Capacity to Implement

在你的亲戚朋友中开展一项调查，从中找到三四个患有头痛的人，问他们现在服用什么药物。如果他们不服药，询问他们用什么方法治疗自己的头痛。列出你收集到的所有治疗头痛的方法。现在向被你调查过的人推荐其他治疗方法。对于使用人工合成药物的人，看看是否能说服他们改用天然方法治疗。

故事灵感来自
This Fable Is Inspired by

珍妮特·多纳
Janet Dohner

　　珍妮特·多纳是一位经验丰富的图书馆员，目前居住在美国密歇根州德克斯特市的一家农场。她是研究看护动物的历史知名品种的专业顾问，对看护犬有超过30年的研究经验，著有《牲畜的守护者：利用狗、驴和羊驼来保护你的牧群》，帮助牲畜养殖企业获得成功。她还著有《历史知名及濒危牲畜家禽品种大全》一书。珍妮特靠销售自己农场生产的设得兰羊毛、棉絮、纱线，以及她的多部著作谋生。

图书在版编目（CIP）数据

冈特生态童书.第四辑：修订版：全36册：汉英对照 /
（比）冈特·鲍利著；（哥伦）凯瑟琳娜·巴赫绘；
何家振等译.—上海：上海远东出版社，2023
书名原文：Gunter's Fables
ISBN 978-7-5476-1931-5

Ⅰ.①冈… Ⅱ.①冈… ②凯… ③何… Ⅲ.①生态环
境−环境保护−儿童读物—汉、英 Ⅳ.①X171.1−49

中国国家版本馆CIP数据核字（2023）第120983号
著作权合同登记号图字09-2023-0612号

策　　划 张　蓉
责任编辑 张君钦
封面设计 魏　来 李　廉

冈特生态童书
快速止痛药
[比]冈特·鲍利　著
[哥伦]凯瑟琳娜·巴赫　绘
章里西　译

记得要和身边的小朋友分享环保知识哦！
八喜冰淇淋祝你成为环保小使者！